INTRODUCTION

This module was written in accordance with the Curriculum and Assessment Standard Document for the subject of Technology Design (RBT). This module strengthens and equips junior readers aged between 4 to 15 years old with the knowledge and skills to learn electronic topics. This module is specially designed for readers with beginner knowledge in programming and microprocessor use.

This book is specially designed for UNO Maker boards. This module can also be used for all Arduino UNO compatible microprocessor boards, such as CT-UNO. This book is usually accompanied by a set of student modules that can be used in conjunction with the tutor module. This module is built and written by teachers, for teachers; in collaboration with GMN Technologies Academy.

GMN Academy is a social enterprise based in Malaysia. The current Academy is founded by 4 high school teachers whose mission is to provide quality and relevant STEM education for the underprivileged. Arduino emphasizes the spirit and principles of Open Source. The capital is also issued on the same principle.

Users are encouraged to share, modify and redistribute the information contained in this module. Users only need to give credit to the author of this module. Don't forget to review after you have read it. So, happy learning kids!!

CONTENT

Preparation
UNIT 1 - Microcontroller
    1.1 WHAT IS MICROCONTROLLER AND ITS FUNTION
    1.2 PARTS IN MICROCONTROLLER
    1.3 BLOCK DIAGRAM
    1.4 INTRODUCTION TO SCHEMATIC DIAGRAM
UNIT 2 - OUTPUT
    2.1 INTRODUCTION TO OUTPUT PROGRAMMING
    2.2 TYPE OF OUTPUT DEVICES
    2.3 INTRODUCTION TO OUTPUT TABLE
    2.4 OUTPUT SIMULATION CIRCUIT
UNIT 3 - INPUT
    3.1 INTRODUCTION TO INPUT PROGRAMMING
    3.2 TYPES OF INPUT DEVICES
    3.3 INTRODUCTION TO INPUT TABLE
    3.4 INPUT TABLE
UNIT 4 –COMBINATION OF OUTPUT AND INPUT
    4.1 INTRODUCTION TO PROGRAMMING OPTIONS
         CONTROL STRUCTURE
    4.2 SWITCHING CIRCUIT INPUT AND OUTPUT
    4.3 SIMULATION CIRCUIT INPUT AND OUTPUT
    4.4 CREATING PROJECT
APPENDIX 1 HOW TO PROGRAMMING TO SMARTPHONES

Preparation

1. **Microcontroller Arduino**
   You can buy Maker UNO set from online store here:

   Link 1 :
   Link 2 :
   Link 3 :

2. **Arduino software**
   Arduino software can be downloaded from the official Arduino website:

   https://www.arduino.cc/en/Main/Software

UNIT 1
WHAT IS MICROCONTROLLER AND MICROPROCESSOR

Microprocessor control device. This control device uses microprocessors as central processing units (CPUs), RAM (Random Access Memory), ROM (Read Only Memory) as well as Input and Output (I/O) devices.

CPU, RAM, ROM and I / O devices are in a single chip. The microprocessor is a central processing unit (CPU) that exists as a single chip. RAM, ROM and I/O device are separated with the CPU.

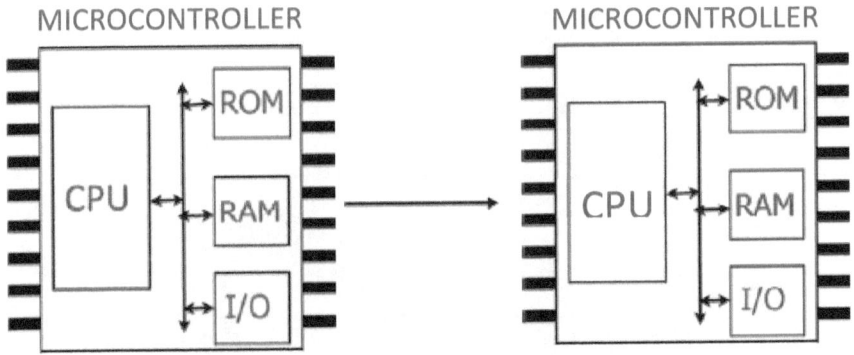

Figure 1.1(A): This figure shows that the CPU is a microprocessor

Microcontrollers are often used in various devices around us such as in televisions, air conditioners, and modern car.

UNIT 1.1

Microcontroller is responsible for receiving input, processing the information received and producing the correct output. A microcontroller can be considered a small computer or a brain.

Comparison can be made with figure 1.1 (b) below. Microcontroller systems act like the human nervous system. The stimulating nerve receives input from the hand and sends that information to the spinal cord for processing inside brain.

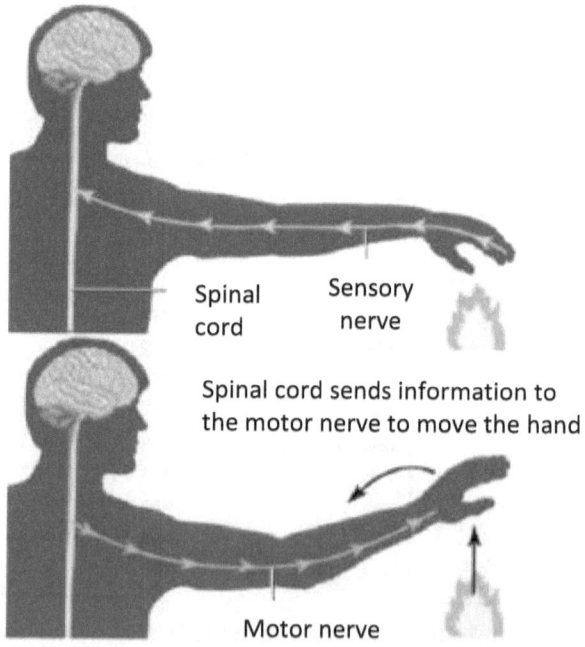

Figure 1.1(B): Human stimulus nerve system

A microprocessor can be instructed through programming to perform tasks automatically. After giving instructions, a microprocessor will save the command without the need for a computer connection. In general, the microprocessor works as in Figure 1.1 (c) which is to obtain information (input) and response (output) according to the instructions given.

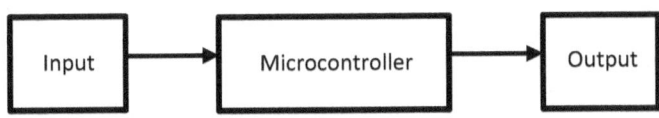

Figure 1.1(C) : Figure of microcontroller block

UNIT 1.1

Figure 1.1(D) : Types of Microcontrollers Microcontroller

Among the most commonly used Microcontroller Development Board are as follows:

Figure 1.1(E): Intel Edison

Figure 1.1(F): Arduino UNO

Figure 1.1(G): Maker UNO

Figure 1.1(H): Beaglebone black

UNIT 1.2

PARTS INSIDE MICROCONTROLLER

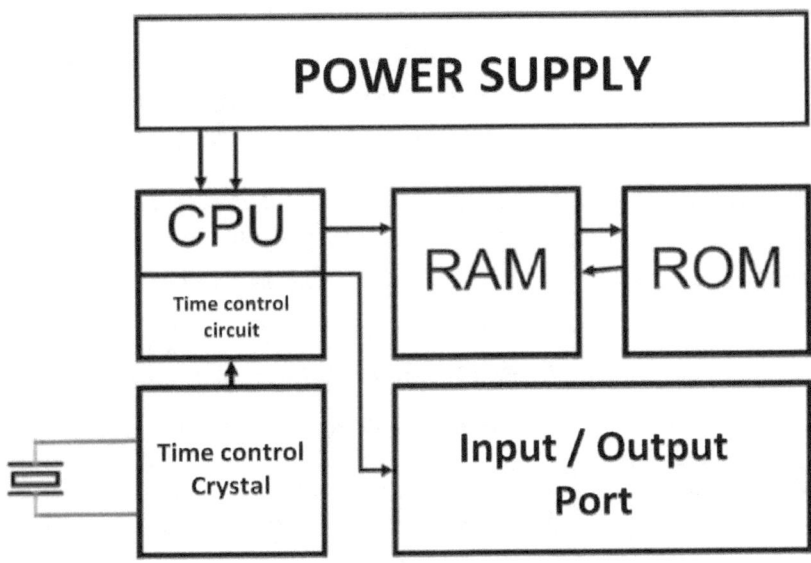

Figure 1.2 (a): The diagram shows the parts in the microprocessor

UNIT 1.2

Parts	Functions
CPU	Obtain information and instructions (program) to process input and output
RAM & ROM	Memory space for storing information and instructions (programs)
Parallel Input/output	Connects to input nodes and output such as LED, motor and sensors. There are two types of Input and Output signals which are analog signals and digital signal
Timer Circuit	Provides microprocessor capability to control system over time.
Crystal Time Guard	Used to produce the frequency used in the time controller circuit.
Power supply	Provides electrical power to microcontroller

UNIT 1.3

INTRODUCTION TO BLOCK DIAGRAM

Figure below show an example of block diagram:

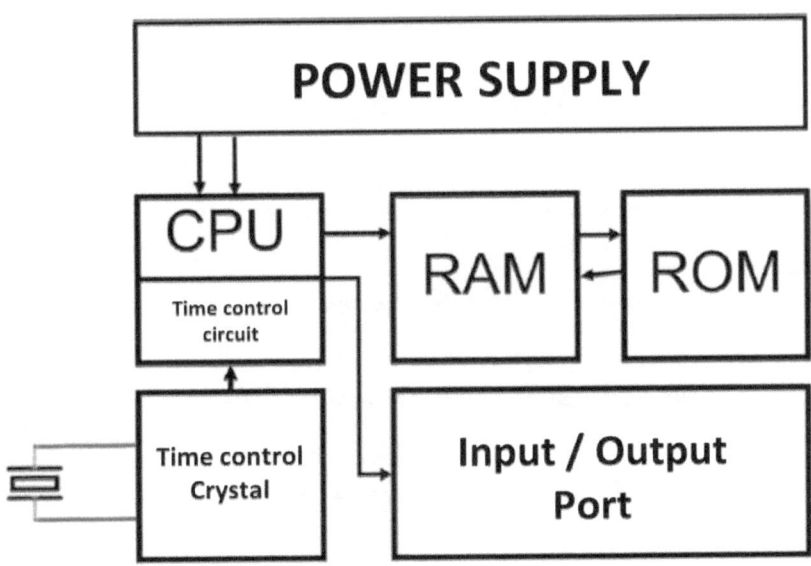

Figure 1.3 (a): The diagram shows the block diagram of microprocessor

UNIT 1.4

INTRODUCTION TO SCHEMATIC DIAGRAM

Guidelines for generating electronic circuit design sketches as below :

1. Make sure the lines are drawn straight

2. Make sure the lines drawn are not arrows.

3. Make sure the lines are vertical or horizontal only.

4. Reduce line shortcuts to avoid confusion

5. Use fixed symbols to describe each component.

6. Make sure each component in the schematic diagram is labelled.

UNIT 1.4

Essential elements in schematic diagram:

Parts		Functions
⏚	Ground (GND)	A negative terminal in a circuit. In general, electronic circuits will end with GND.
5V ⊥	Power supply, 5V	Electronic circuit start up. Source of power supply

The following is a schematic diagram of an electronic circuit.

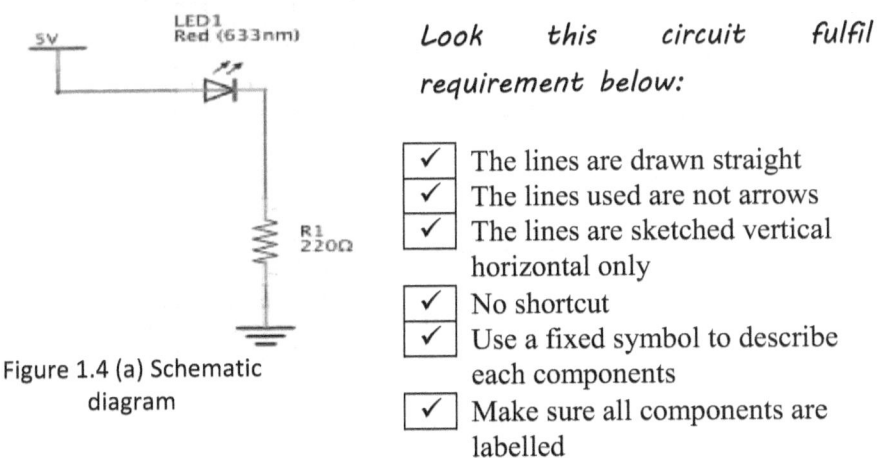

Figure 1.4 (a) Schematic diagram

Look this circuit fulfil requirement below:

- ✓ The lines are drawn straight
- ✓ The lines used are not arrows
- ✓ The lines are sketched vertical horizontal only
- ✓ No shortcut
- ✓ Use a fixed symbol to describe each components
- ✓ Make sure all components are labelled

Here is the symbol for the microcontroller:

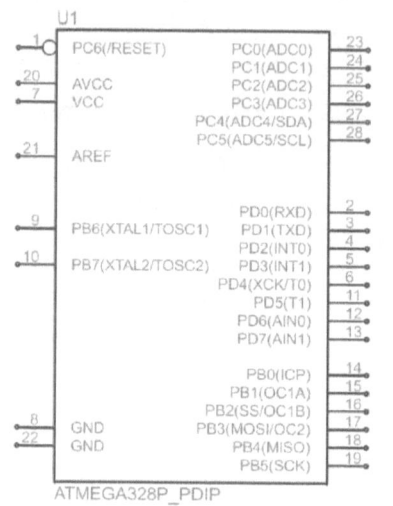

Figure 1.4 (B): Schematic diagram for ATMEGA328P chip (microphone controller in UNO Arduino board)

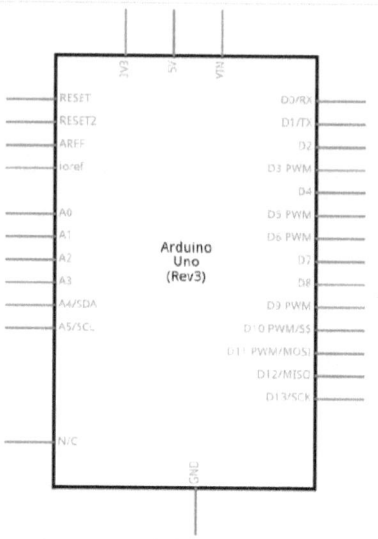

Figure 1.4 (C): Schematic diagram of the Arduino UNO microcontroller board

The following is an example of a schematic diagram that use microcontroller

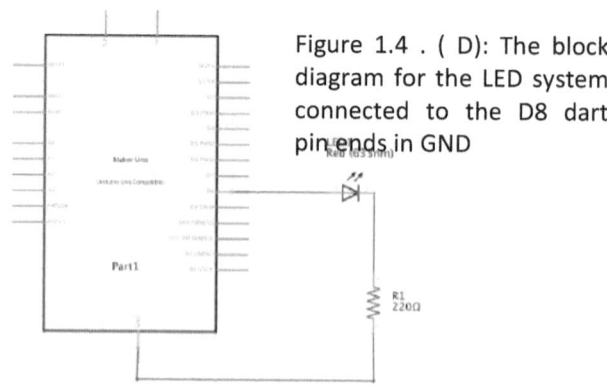

Figure 1.4 . (D): The block diagram for the LED system connected to the D8 dart pin ends in GND

UNIT 2
INTRODUCTION TO OUTPUT PROGRAMMING

The microprocessor can function to control the output circuit by sending a signal to the output pin on the microprocessor. Thus, the electronic components connected to the microprocessor pin can be controlled through programming.

In general, there are 2 types of signals that can be emitted from the microcontroller, namely digital signal and analog signal. Before beginning to write a program, we need to understand the difference between digital signal and analog signal.

In short, digital signals have two states which are 19N (circuit with power current) or OFF (no current in the circuit sends digital signal ON to pin 5, we can activate the connected component at pin 5.

While if we send the digital signal OFF to pin 5, then the component connected to pin 5 will be deactivated. For analogue signal, it has more than 2 values, and the Arduino microphone controller, is comprised of 256 values ranging from 0 to 255.

Value 0 means Electric current voltage (voltage value 0) and value 255 is maximum value. The value of the analog signal (0 to 255) is directly proportional to the voltage produced on the output pin (for example the digital signal 127 will generate 50% of the maximum voltage).

For easier understanding, if we connect LED on pin,: digital and analog output voltage, value 0 will turn off LED, value 127 makes 1-lit LED with 50% brightness, n value 255 will turn on the LED with 100% brightness.

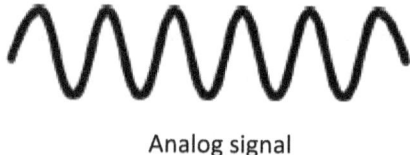

Analog signal

Digital signal

Figure 2.1 (a). Comparison between analog signal and digital signal

The Arduino microprocessor performs two main functions, namely the Setup function and the Function loop. The setup function will be executed only once the Arduino is turned on or reset, while the Loop function will be repeated several times until the Arduino is deleted.

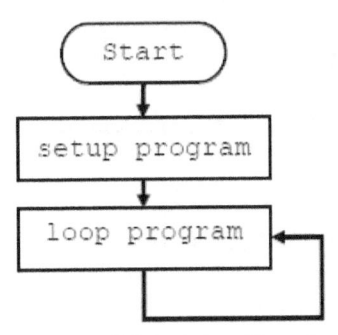

Figure 2.1 (b) Arduino Flowchart

The program can be written on the Arduino IDE (Integrated Development Environment) software development program only for Arduino. The software can be downloaded from the official Arduino website as below https://www.arduino.cc

After opening the Arduino IDE software, you are encouraged to display a line number to facilitate programming. Click on File - Preferences, mark it on the Display Line Number. Click OK.

You will find that the Arduino software interface is as in Fig. 2.1 (c). Note the two Functions that have been automatically built into the Arduino sketch - setup and loop.

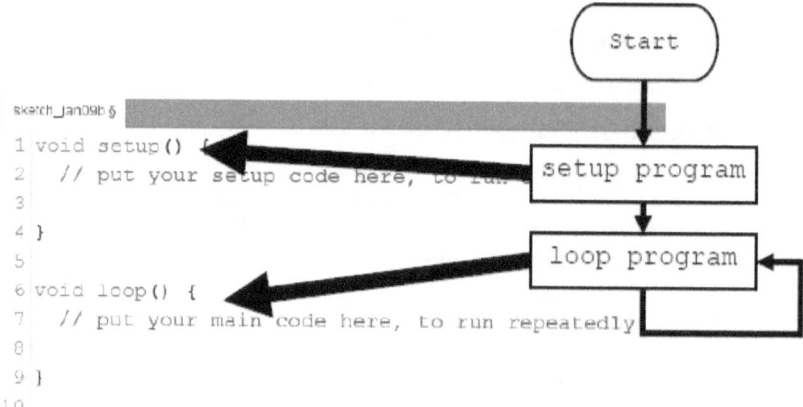

Figure 2.1. C): Comparison between a flow chart of the Arduino with the program how the Arduino IDE

UNIT 2.1

Steps to Arduino programming:

Step 1 — Set pinMode at void setup()
Identify the pin number to use with the pin mode, either input (output) or output mode.

Step 2 — Design your program
Think about what processes you need to achieve with your programming. Make a flow chart if needed.

Step 3 — Write the program in the function of void loop ()
Write a program in the loop function. Your program could be involved in output signal transmission, reception, reading the input signal, and so on. The main rules of Arduino programming:

Make sure that each program line ends with a mark; (semicolon).
Make sure the keyword spelling (keyword) is correct (orange and blue).
Make sure each sign () and {} are in pair.

Steps-to writing an Arduino program:

Step 4 — Connect the Arduino board to your computer using USB cable

Connect the Arduino board to your computer using a USB cable, and make sure the correct Port is selected.

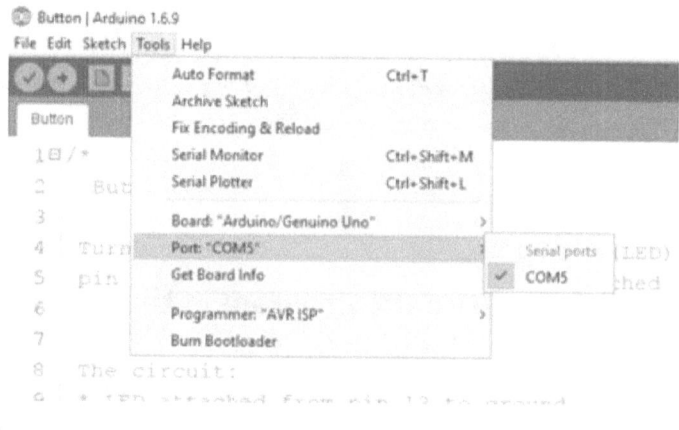

Step 5 — Click Upload icon (Right arrow) and save the file

Once the code has been uploaded to the Arduino, the command code will remain in the Arduino memory until the program has been erased or rewritten. Before we write a program, we need to plan a programs by using a flow chart.

Here are the set of steps for programs to be developed.

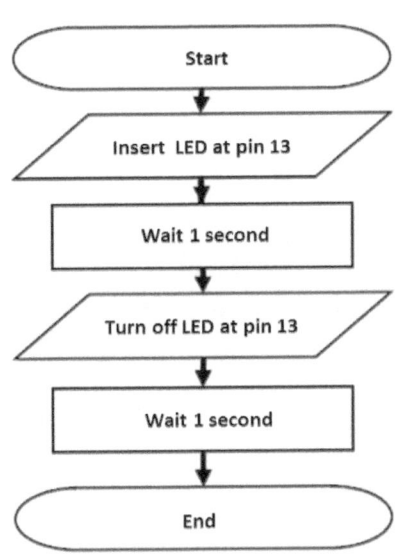

This is an example of a sequence control structure - a non-branching program. Note the different shapes used to represent a different program

⬭ Represents the start and finish of a program.

▱ Represents the input or output of a program.

▭ Represents the steps that need to be processed in a programs.

21

Following below is a programs to the flow chart from previous page.

```
1  void setup() {
2     pinMode(13,OUTPUT);
3  }
4
5  void loop() {
6     digitalWrite(13,HIGH);
7     delay(1000);
8     digitalWrite(13,LOW);
9     delay(1000);
10 }
```

The functions for each row are as follows:

Line	Commands	Functions
1	void setup() {	{Declaring the start of the setup functions. Setup programs are typically used for basic settings such as setting the mode for the pin used. Note that the words void and setup are in different colours, as they are a special keyword in Arduino program-ming. The symbol {marks the begin-ning of the setup program.

2	pinMode (13, OUTPUT);	Set a specific pin mode to either input mode or output mode This code sets pin 13 as an output mode.
3	}	Symbols } indicates the end of the setup function. Starting with the symbol { in line 1, the setup function is only includes code between rows 1 and 3.
5	void loop () {	Indicates the beginning for the loop function. The loop function is a function that is repeated until the Arduino is deleted. Take note of the colours in the words void and loop, both of which are different - just like the setup of the word loop is the key word in Arduino. The symbol { marks the beginning of the loop function
6	digitalWrite(13,HIGH);	DigitalWrite sends a digital signal to the specified pin. This line will send the HIGH value to pin 13, which will turn on the components connected to pin 13. Arduino is sensitive towards uppercase and lowercase letter - W for digitalWrite and HIGH should be written in new capital letters, then it will be identified as a keywords and will be displayed in other colours

7	delay(1000)	Delay will wait for the specified time. The value expressed in parentheses is in milliseconds. For example, 1000 milliseconds equals to 1 second. So delay(1000) means the system will wait for 1 second.
8	digitalWrite(13,LOW);	This line will send the LOW value to pin 13. This will erase / deactivate the components connected to pin 13.
9	delay(1000)	We need to place another delay after deleting pin 13. Without this row, after row 8 is run, the program flow will return to line 6. Although pin 13 will be deleted, it will delete so fast and it is out of sight of our rough eyes, so it seems like it is never deleted.
10	}	Symbol } marks the end of the loop function. Starting with the symbol { in line 5, the loop function only includes code between rows 5 and 10.

With this, we can control pin 13 to turn it on and off at any time. You will notice that the Arduino board's LEDs are flashing. This is because pin 13 is also the pin of LED in

You can also use analogWrite to send signals to pins on the Arduino. Not all Arduino pins can emit analog signals, only pins have symbols ~ only on Arduino boards can emit analog signals, namely 9, 10, 11, 3, 5, 6 and 7 pins.

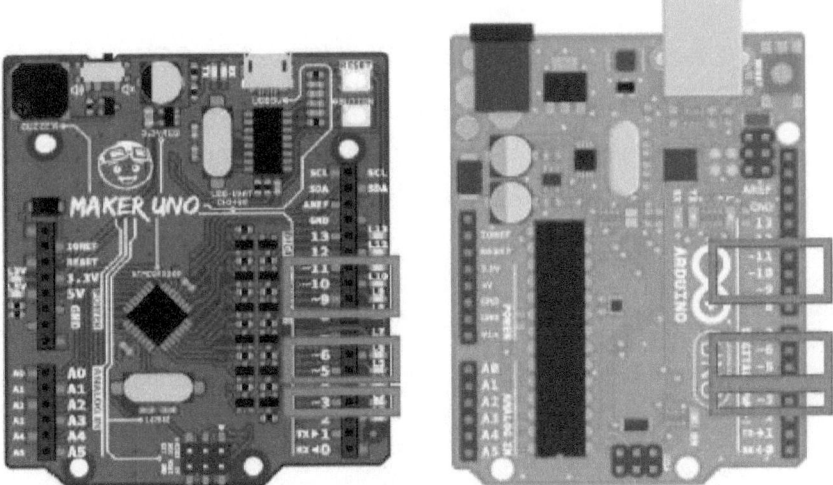

Figure 2.1 (d): Location of output analog pin also known as PWM pin

The instructions for analogWrite are as follows:

analogWrite**(pin, value)** where the value ranges from 0 to 255

Example:
- analogWrite(9,255) *will send the value 255 which is the maximum voltage to pin 9. Any component connected to pin 9 will receive the maximum voltage. If the LED is on pin 9, it will light up with 100% brightness.*

- analogWrite(9,127) *will send the value 127 (i.e. the maximum half voltage to pin 9). Any component connected to pin 9 will receive a voltage of half its maximum and the LED mounted on that pin will light up with 50%*

- analogWrite(9, 0) *will send the value 0 (i.e. voltage 0) to pin 9. Distribute the component connected to pin 9 will not receive the current because the LED volt connected to pin 9 will turned off .*

The following sections are for UNO Maker only. For other Arduino, the LED needs to be connected to the circuit.

Maker UNO has LEDs on every digital pin. You can write programs how to control LEDs on each pin. (See next page for example).

For example:

```
digitalWrite( 9, LOW);
digitalWrite(8, HIGH);
delay (1000);
digitalWrite(9, HIGH);
digitalWrite(8, LOW);
delay (1000);
```

This setup will allow LED to blink on pin 8 and 9 simultaneously.

UNIT 2.2
TYPE OF OUTPUT DEVICES

The output device will execute the signal provided by the pin on the Arduino. Arduino can supply 5v voltage to power devices connected to the Arduino. The most commonly used output devices are light emitting diodes and sensors. This output device can be powered on without additional power supply. There are also other output devices that require additional power supply such as 12v direct current motors, 12v LED lights and so on.

Resistor

1. Resistors are used to limit the current flow. The higher the resistor value, the higher resistance for current flow through it.
2. Resistor is important to prevent the devices from being damaged or as a voltage divider
3. Resistor value can be read by looking at the colour of the strip on resistor.
4. The schematic diagram is ─/\/\/\─

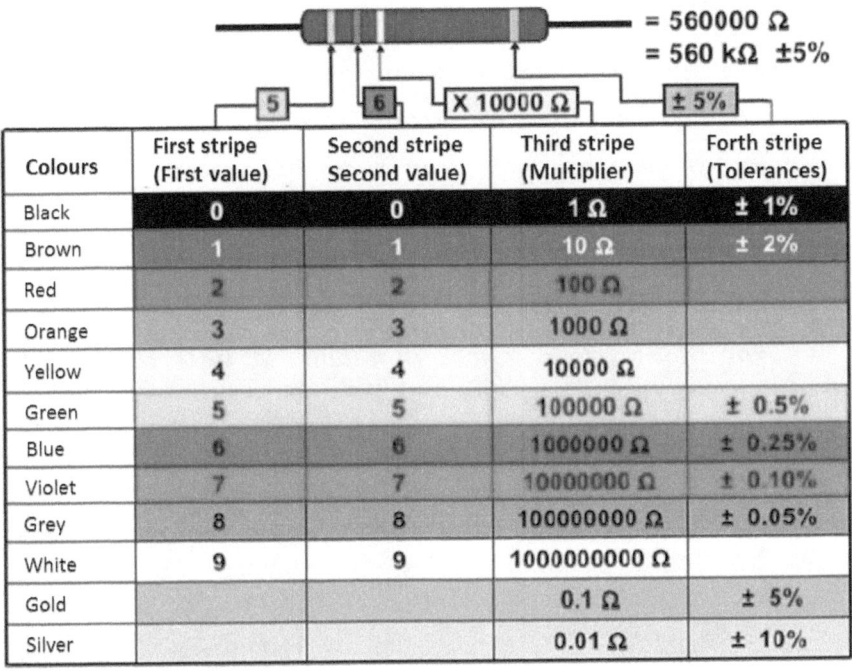

Colours	First stripe (First value)	Second stripe (Second value)	Third stripe (Multiplier)	Forth stripe (Tolerances)
Black	0	0	1 Ω	± 1%
Brown	1	1	10 Ω	± 2%
Red	2	2	100 Ω	
Orange	3	3	1000 Ω	
Yellow	4	4	10000 Ω	
Green	5	5	100000 Ω	± 0.5%
Blue	6	6	1000000 Ω	± 0.25%
Violet	7	7	10000000 Ω	± 0.10%
Grey	8	8	100000000 Ω	± 0.05%
White	9	9	1000000000 Ω	
Gold			0.1 Ω	± 5%
Silver			0.01 Ω	± 10%

Figure 2.2 (a): How to read the value of resistor

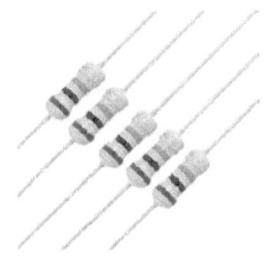

Figure 2.2 (b): 10,000 Ohm resistor

Light Emitting Diode (LED)

- Light emitting diode is a diode that emits light when the current passes through it.
- LEDs can be used as decorative lamps and should be used in conjunction with resistors to prevent them from burning.
- The LED has polarity on its connectivity, which is positive and negative.
- For Maker UNO, there is a built-in LED on each digital pin
- LED symbol is :

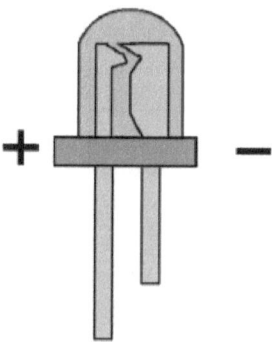

Figure 2.2 (c): How to identify positive and negative feet for LEDs

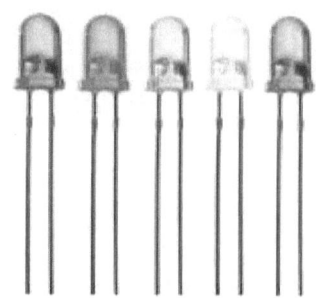

Figure 2.2 . (D): LED image

Buzzer

- *Buzzer function is converting electrical signal to sound.*
- *Like LEDs, buzzer also has polarity on the connection.*
- *Buzzer can either be used as a security alarm or play a musical melody.*
- *The symbol for the pointer is :*

Figure 2.2 (e): Different types of buzzer

On Maker UNO, there is a built-in speaker (built in buzzer on Maker UNO connected to pin 8 via sliding switch. The pointer must be activated using the slide switch before it can be used.

Sliding switch

Buzzer

Figure 2.2. (F): The location of the buzzer and the switch, on UNO Maker

31

Direct Current Motor

- There are many types of motors that require different voltages, for example 3v, 6v, 9v, 12v and others.
- The current motor shaft will rotate when there is an electric current pass through it. Generally, the motor needs to be connected to the driver, or to the transistor and the diode as it needs higher current
- Driver (the 3'1g driver is mounted on the motor allowing us to control the direction of the motor either clockwise.) or anti-clockwise.
- For Maker UNO, a 3v direct current motor is provided inside the package. The UNO Maker cannot power a motor that requires more than 5V.
- Motor can be used in conjunction with a transistor because the Arduino pin is unable to generate enough current to power the motor.
- Symbol of C Motor : −[M]−

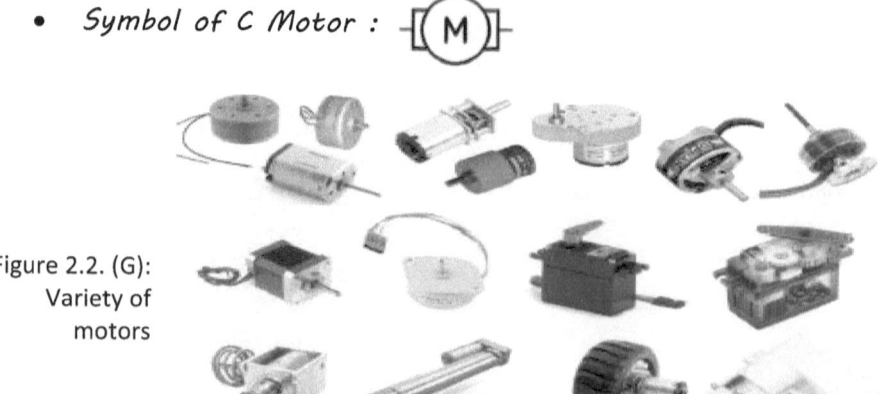

Figure 2.2. (G): Variety of motors

UNIT 2.3
INTRODUCTION TO OUTPUT CIRCUIT CONNECTIVITY

Connection of Output circuits by using jumper wires and breadboard. The dashboard is an electronic board that allows us to test the current circuit using jumper wires that do not require soldering.

Before we start building, we need to understand the functions of the following devices:

Jumper Wire

Jumper wires are used to connect various devices to the microcontroller. Unofficially, a red jumper wire is used for wires connecting to the power supply and black jumper wires are used for wires connecting to the earth.

Figure 2.3 (a): Picture of jumper wire

Circuit Board

- Breadboard is used to test and produce prototype circuits without soldering.
- The holes have a connecting system as shown in Fig. 2.3 (a). Holes in the design board are intersect along the red line. The components or wires to be connected when they are on the same line.

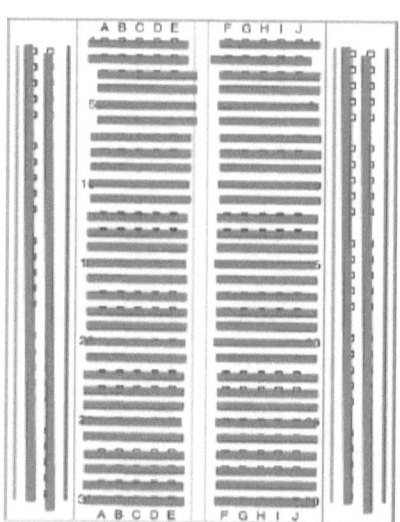

Figure 2.3 (a): The red dotted line shows the connection of board line

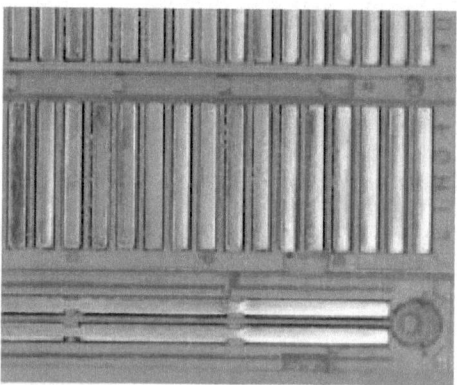

Figure 2.3 (b): The inside of the board — note that the iron column is connecting the circuit

Connection of output circuit to circuit board

LED connectivity

The following is an example of connecting a series of circuits and simple parallel circuits using a 3V battery as well as a design board. The diagrams show schematic diagrams as well as connections to the circuit board.

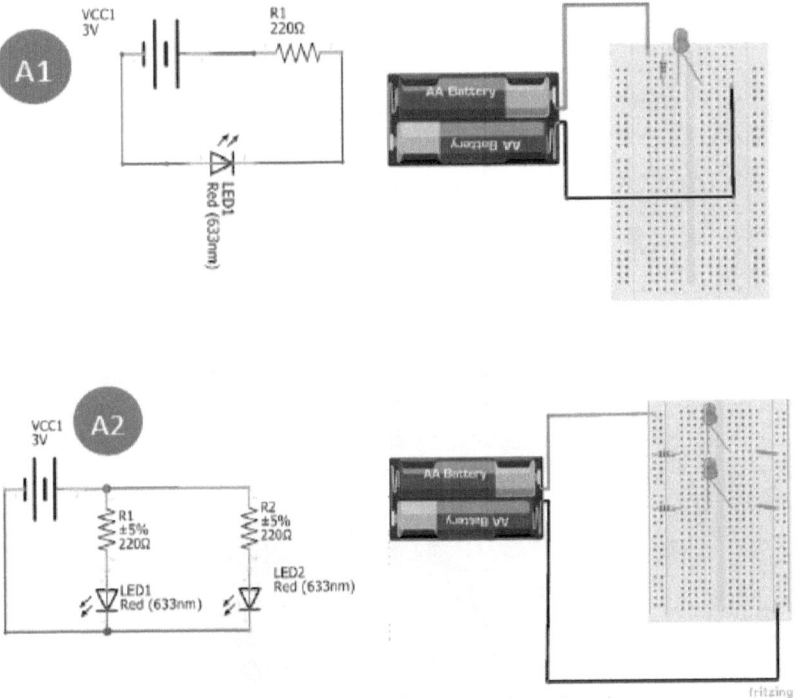

Figure 2.2 (A): A1 Circuit and A2 (Top)
Circuit in series: (Bottom) Parallel circuit

Imagine this, the battery connection is replaced with a 5v current source.

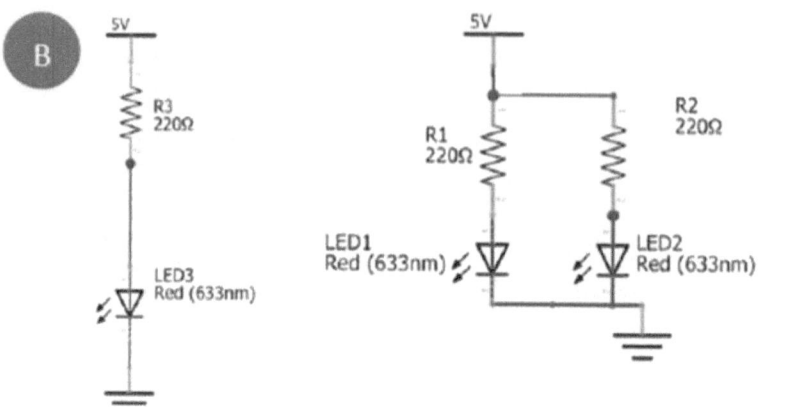

The Arduino or UNO Maker can also supply 5v voltage sources such as 5v pin battery as well as GND pin (Earth).

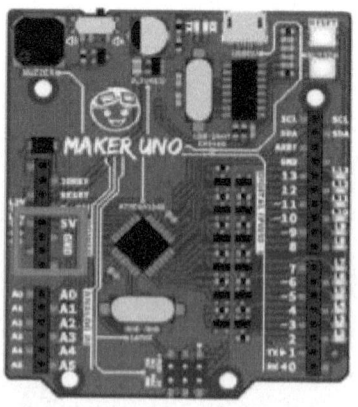

Buzzer

The buzzer should be connected in polarity like the LED (red to pin and black to ground). Here is an example of the connectivity of the buzzer on Arduino. For the UNO Maker model, the built-in buzzer are provided on pin 8. The buzzer needs to be activated by pushing the slide switch before use.

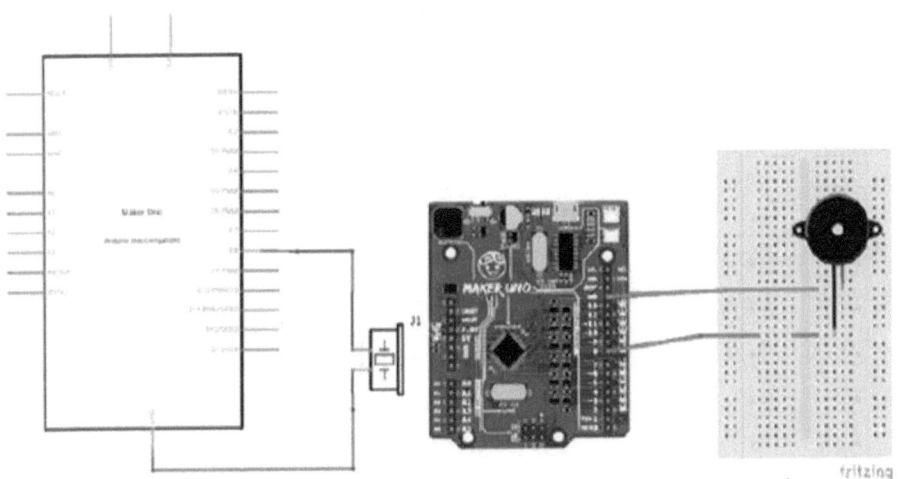

Programming for Buzzer

```
1   void setup () {
2       //put your setup context
3       pinMode(8, OUTPUT);
4   }
5
6   void loop() {
7       // put your main context
8       tone(8, 165, 500);
9       delay(1000);
10      tone(8. 175, 500);
11      delay(1000);
12      tone(8. 196, 500);
13      delay(1000);
14  }
```

Line	Programming	Functions
8	tone(8, 165, 500);	This line will play 165Hz sound, which is "63" in music at pin 8 for 500 milliseconds. Tone is a special function used to control buzzer. It take 3 information namely the pin number, frequency of sound to be played and the duration of time the sound is played.

Tone is capable of producing frequencies equal to music notes. Here is a list of music notes (in octave 3) as well as their frequency.

Music Note	Frequency(Hz)
E	165
F	175
G	196
A	220
B	247
C	262
D	294
E	330
F	349

Direct Current Motor

DC motor needs to be connected to the transistor. This is because the motor requires a high current to operate and it cannot be supplied by the Arduino Chest output pin. Its circular connection is as the following:

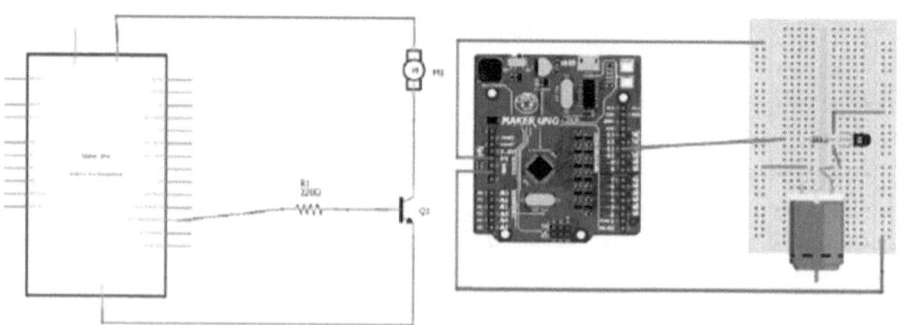

To solve this problem, we need a transistor that will act as a switch to control the motor. To activate the motor, we just need to send a signal to pin 11 and then it will "open the transistor to allow the current to flow to the motor.

Programs for starting the motor are as follows:

```
1  void setup () {
2      //put your setup code here
3      pinMode(11, OUTPUT);
4  }
5
6  void loop() {
7      // put your main context
8      digitalWrite(11, HIGH);
9  }
```

UNIT 2.4
OUTPUT CIRCUIT SIMULATIONS

To produce output circuit simulation, we can use the TINKERCAD website for free.

Go to website tinkercad : www.tinkercad.com

Here is the list of steps to gain access to tinkerlab

1. Sign up for an account with tinkercad

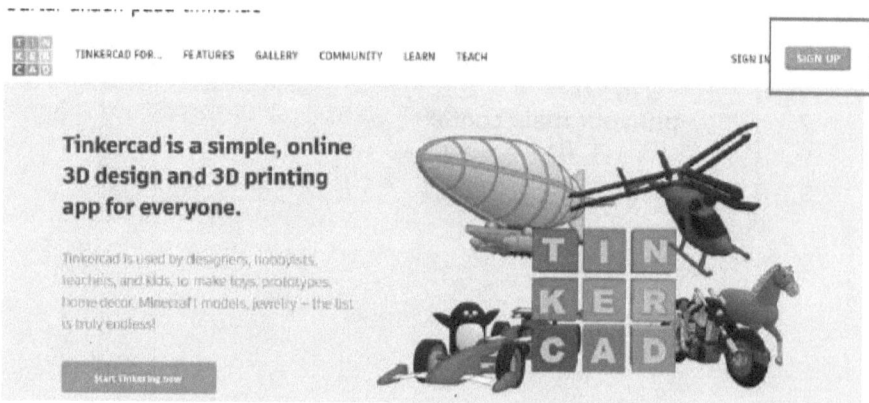

2. Log in to your account and click "Circuits"

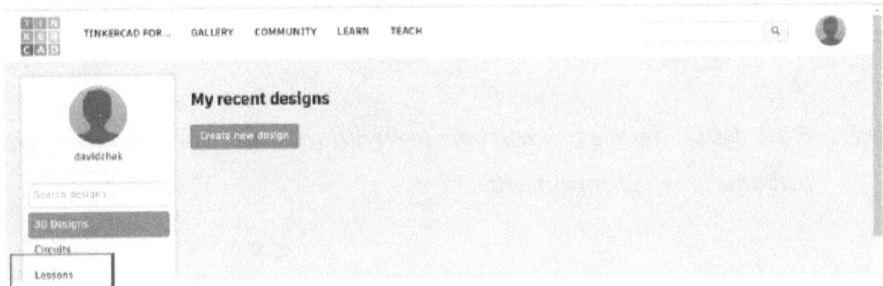

3. Click "Create New Circuit"

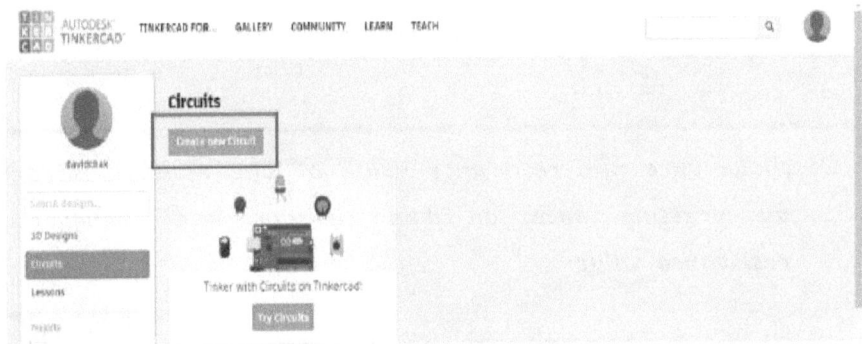

43

The making of simulation output of 1 LED circuit connected to pin 5

1. Add LED devices, resistors, Arduino and even boards with butane a + Components "

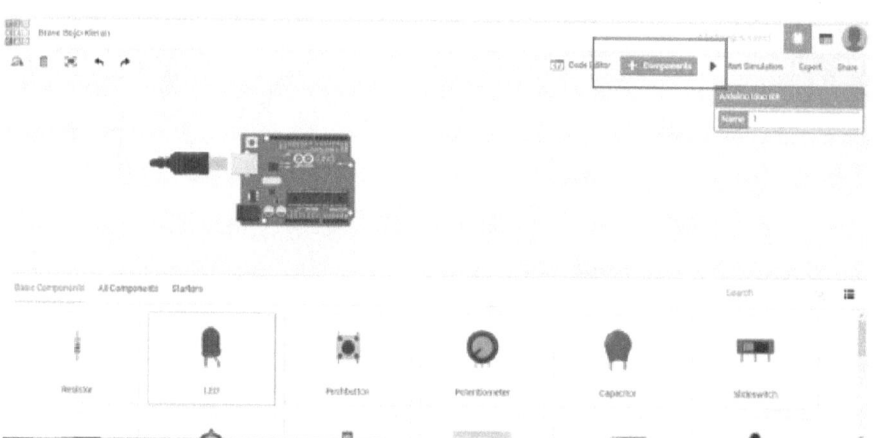

2. Make sure the resistance value of the resistor is correct by pressing once on the resistor and changing the resistance value.

3. Connect the devices on circuit board using the technique of "rap and drop"

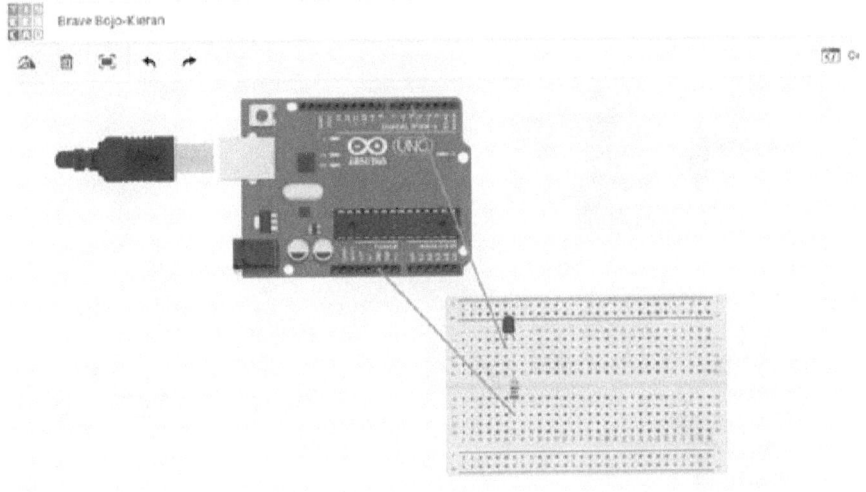

4. Write a program in section "Code Editor"

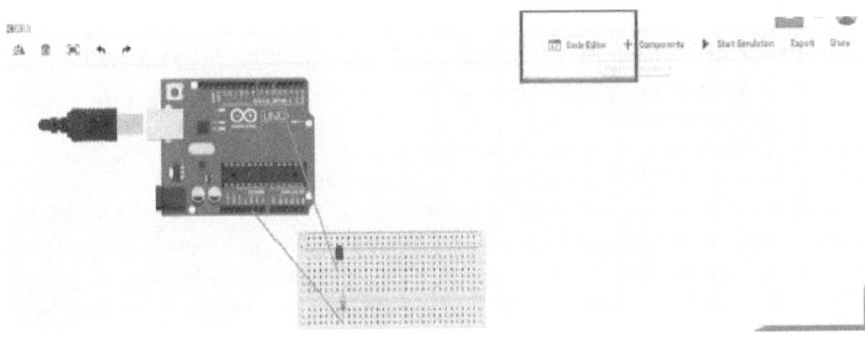

5. "Drag and Drop" the appropriate block for generating a program to allow the LED to flash on pin 5.

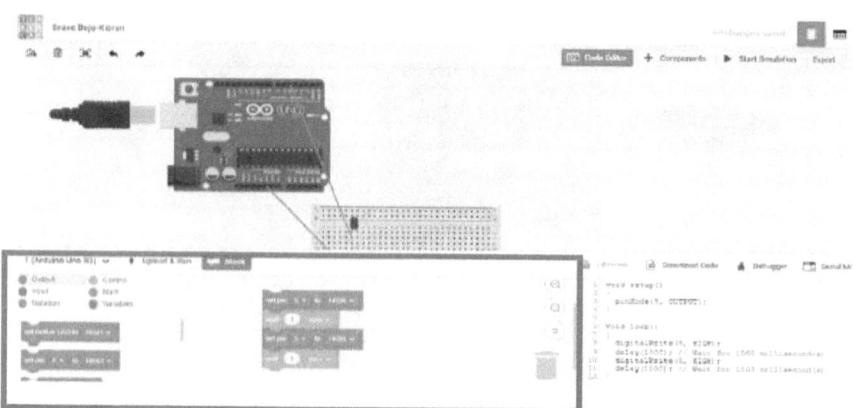

6. After the program I was completely prepared, press upload and Run.

Try to produce simulated output circuit of more than 1 LED, speaker or motor. For more detailed guidance you can refer to the video guide or step-by-step guide.

Steps to access the TinkerCad usage guide

1. Sign in to your account and press "LEARN"

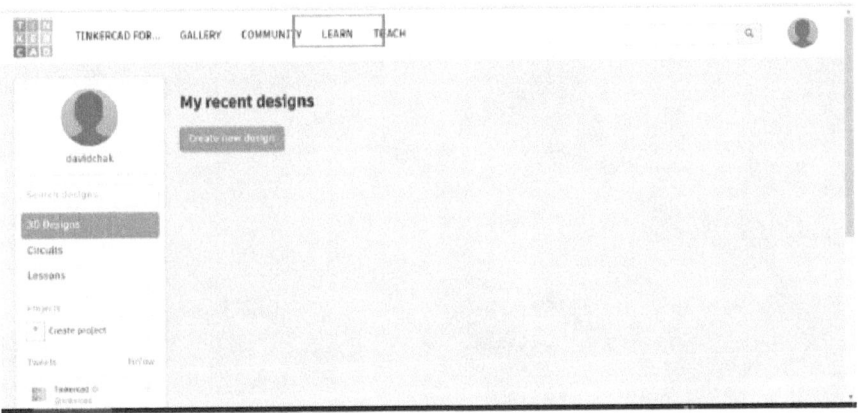

2. Select "Circuits"

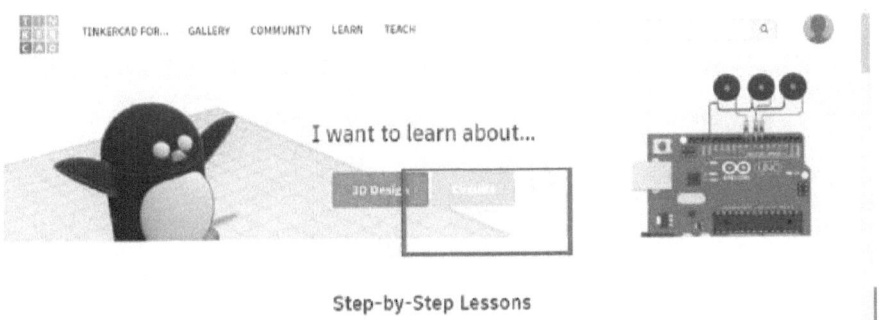

3. You can follow the step-by-step lesson to learn how to use Tinker or theories about the Arduino and its electronic components

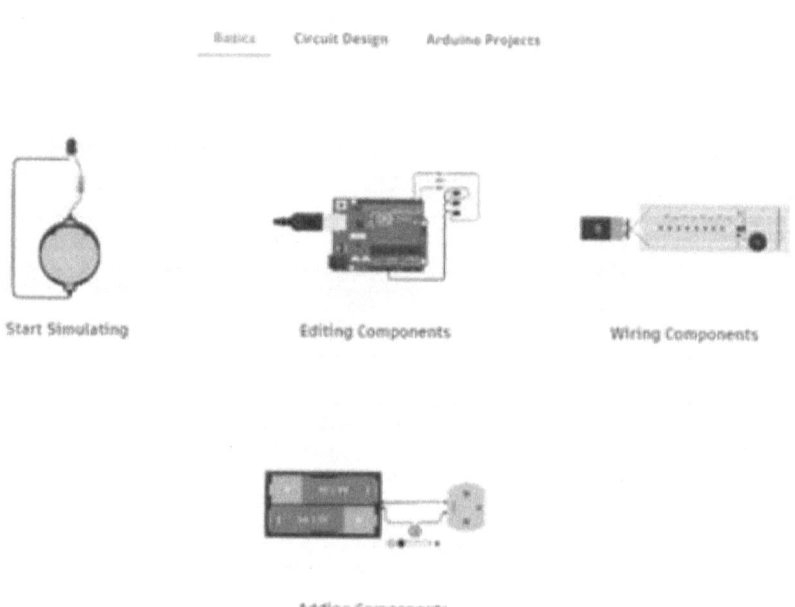

4. You can also learn through video tutorials.

Video Tutorials

Pushbutton

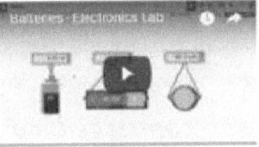
Batteries

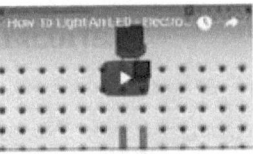
Light an LED

See all videos

Circuits Keyboard Shortcuts

General shortcuts

ctrl + C	Copy object(s)
ctrl + X	Cut object(s)
ctrl + V	Paste object(s)
ctrl + Z	Undo action(s)
ctrl + Y	Re-do action(s)
ctrl + D	Deselect
ctrl + I	Invert selection

Moving object(s)

↑ / ← / ↓ / →	Move one unit
shift + ↑ / ← / ↓ / →	Move 1/10 unit
R	Rotate Clockwise
shift + R	Rotate Counterclockwise

Keyboard + mouse shortcuts (press and hold keyboard button, then move mouse)

shift + left mouse button — Select multiple object(s)

49

UNIT 3
INPUT

INTRODUCTION TO INPUT CIRCUIT PROGRAMMING

The microprocessor can receive a signal via pin on the microprocessor. The sensor connected to the microcontroller can transmit digital signals or analog signals.

Digital signals have two kinds of value that is equal 1 (On) or 0 (Off). All pin numbers (3-13) can receive digital signals through the code "digitalRead". An example of an input device that uses a digital signal is a device that has two conditions, such as a switch (pressed / non-pressed switch). rain sensors (has water / no water) and so on.

Analog signal contains more than two values, which range from 0 to 1023. Analog signal can only be read through analog pin A0 to A5 via "analogRead" code. Examples of input devices that use analog signals are devices that detect more than two conditions, such as adjustable resistors (values change according to knob rotation), light sensors (very dark to very bright). microphone and so on.

Before we use any input device, we need to set the pin as INPUT in the setup section like this:

```
sketch_jan07a §
1  void setup() {
2      // put your setup code h
3      pinMode(4, INPUT);
4  }
5
```

Replace 4 with pin value used in circuits

For Maker UNO, there is a one switch, built in to pin 2. In order to use it, we need to set it as INPUT_PULLUP

```
sketch_jan07a §
1  void setup() {
2      // put your setup code here,
3      pinMode(2, INPUT_PULLUP);
4  }
5
```

51

To use the read values, we need to write a program using analogRead or digitalRead and store that value in a variable like this:

```
6 void loop() {
7     // put your main code here,
8     int x = digitalRead(2);
9     int y = analogRead(A2);
10 }
```

Line 8 reads the digital signal value from pin 2 and stores it in the x variable while row 9 reads the analog signal value from pin A2 and stores it in the y variable.

We can also open serial communications with computers so that we can obtain display values for project creation purposes. To do this, we need to open the channel between the Arduino and the computer in the setup section as follows:

```
1 void setup() {
2     // put your setup code
3     pinMode(A0, INPUT);
4     Serial.begin(9600);
5 }
```

The 9600 is a baudrate of data transmission from the Arduino to a computer. The higher the value, the more data you can send from the Arduino to your computer. The most commonly used brake is 9600.

Next, we can display the read value on a computer using Serial.println (l is small L)

```
void setup() {

   pinMode(A0, INPUT);
   Serial.begin(9600);
}

void loop() {

   int x = analogRead(A2);
   Serial.println(x);
}
```

Once the code has been uploaded to Arduino, we can open the Serial Monitor (CTRL + SHIFT + M) or press the following button to see what Arduino reads.

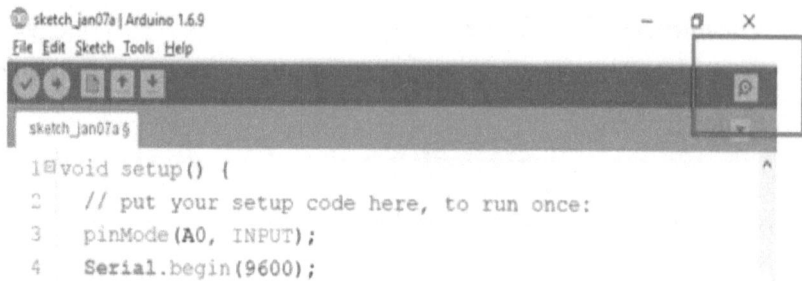

The complete example code below is to read the switch connected to pin 2 (for UNO Maker), and display the value. This example does not require a connection to the design board for Maker UNO

For another Arduino, a circuit board is required. the switch is connected to pin 2 and the setup part must be converted to pinMode(2, INPUT)

Please refer next pages for details understanding.

```
sketch_jan07a§
1  void setup() {
2    // put your setup code here,
3    pinMode(2, INPUT_PULLUP);
4    Serial.begin(9600);
5  }
6
7  void loop() {
8    // put your main code here,
9    int x = digitalRead(2);
10   Serial.println(x);
11 }
12
13
```

UNIT 3.1
TYPES OF INPUT DEVICES

The input device will read the values detected from the environment and send the information read via the pin to the Arduino. All pins can be used to read digital signal values through digitalRead code.

Analog signal can only be detected by using the analog pin A0 to A5. There are various input devices that can be used, such as switches, adjustable resistors, light sensitive or light sensors, temperature sensors, human sensors (PIR Sensors), and so on.

Switch

- There are various types of switches available such as micro-switch, toggle switches and others.
- Although there are many types of switches available in the market, all switches work on the same principle, whether or not the circuit is complete.

- The symbols and pictures for the switch are as follows:

    1. Switch Pushbutton (Close press type) This switch will switch the setting to "ON" while it is pressed. When released, it returns to the "OFF" state.

    Figure 3.1 (a): Image of a pushbutton switch

    Figure 3.1 (b): Schematic diagram of pushbutton switch

    2. Toggle Switch (SPST Type Switch - Single Pole, Single Throw)

    The SPST type switch has two terminals, and the connection can determine whether the circuit is complete or disconnected. SPST switch type connections require pull-up / pull-down circuits for better reading. It will remain "ON" or "OFF" depending on the location of the button.

3. SPDT type switch - Single Pole, Double Throw

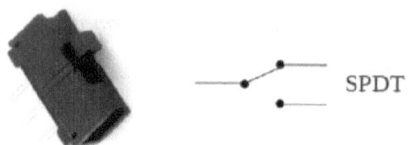

The SPDT type switch has 3 terminals, namely, common terminal, terminal NO (Normally Open) and NC (Normally Close). SPDT type switch connections do not require pull-up / pull-down circuits. It will remain ON or OFF depending on the location of the button.

4. Microswitch is a switch that acts with little movement or light pressure on the switch. An example of microswitch that exist in our daily lives is the light switch in the fridge. When the refrigerator is closed, the light in the refrigerator is turned off.

Adjustable Resistors / Potentiometer

- Adjustable resistors is a type of resistor that can adjust the resistance of the resistor value.
- Resistance value can be changed by rotating the adjust button.
- These resistance changes can be read by Arduino.

Figure 3.1 (C): Adjustable Resistor image

Figure 3.1 (D): Schematic symbol of adjustable resistor

Light Sensor / Light Dependent Resistor

Light-sensitive resistor changes according to the brightness of the light - the brighter, the higher the value that can be read by Arduino.

Figure 3.1 (e): Light-sensitive resistor image

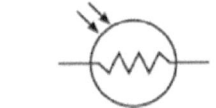
Figure 3.1 (f): Schematic symbols of light-sensitive resistor

Temperature Sensor

- *There are various types of temperature sensors on the market. The most commonly used temperature sensors are the LM35 models.*
- *The LM35 results are in the form of voltage readings, to get results in the form of Celsius, a more in-deep calculation is required.*

Figure 3.1 (g): Temperature sensor image

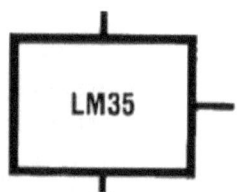

Figure 3.1 (h): Schematic symbols for sensors temperature. This symbol is an example of an IC (integrated circuit)

Passive Movement Sensor / Passive Infrared(PIR Sensor)

- *Human movement can be detected by PIR (Passive Infrared) sensors using infrared light.*

Figure 3.1 (e): PIR sensor

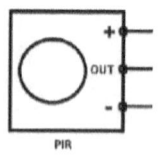

Figure 3.1 (f): Schematic symbols PIR sensor

UNIT 3.3
INTRODUCTION TO INPUT CIRCUIT CONNECTIVITY

Like the output circuit, connection of circuits is by using jumper wires and circuit boards.

Push-Button switch

Maker UNO has a pushbutton switch on pin 2. To use the switch, we need to set pinMode as INPUT_PULLUP in setup section. The connection of pushbutton switch on the board is as follows:

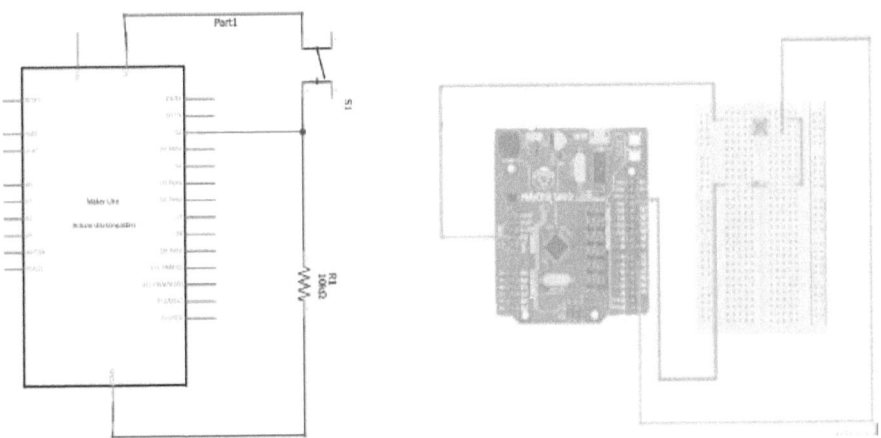

The switch is a digital type device, so we need to use the "digitalRead" code to read the value of the switch.

10k resistor is required to set up PULL DOWN circuit ensuring that the value read is appropriate to attract (pull) all electrical charge to earth 'when the switch is not pressed.

Adjustable Resistor

- Adjustable resistor connections do not have polarity as they are generally resistor.

- Adjustable left and right end resistor pins are connected to 5v and earth while the centre pin must be connected to the pin to be used.

- Adjustable resistors are analog type input devices, so we need to use A0 to A5 pins and use analogRead code commands to read adjustable resistor values. (See next page).

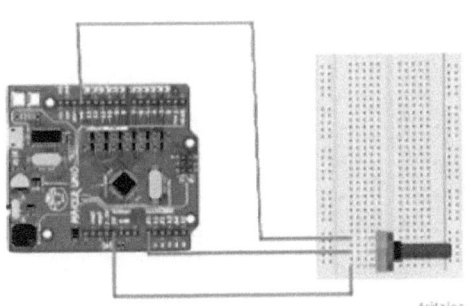
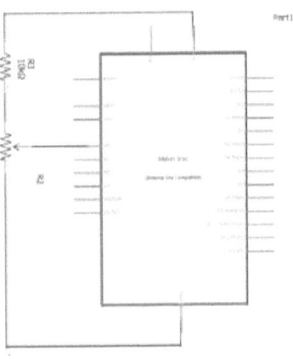

Light Sensitive Resistor

- Like adjustable resistors, light sensitive resistors also have no polarity.
- As with the switch connection, the light sensitive resistor requires a PULL DOWN circuit to obtain readings.
- Light sensitive sensors are analog input devices, so we need to use the A0 - A5 pins and use the analogRead code command to read the resistor values.

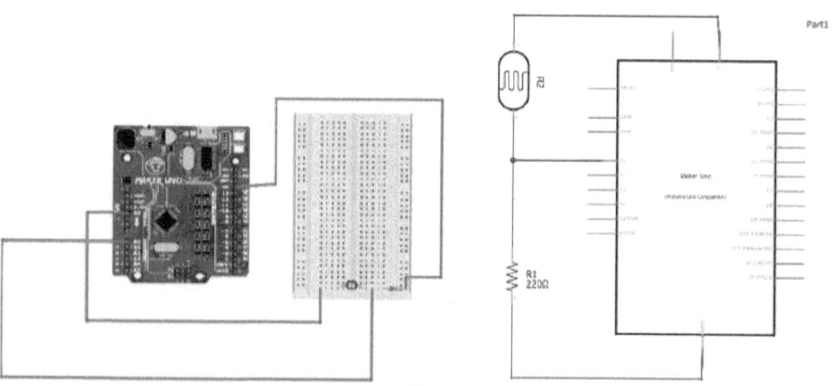

Temperature Sensor

- The LM35 temperature sensor connection is like an adjustable resistor connection where between 3 terminals, the middle end is connected to a pin. The left and right terminals are connected to 5v and Earth.

- The temperature sensor is an analog type input device, so we need to use A0 -A5 pins and use the analogRead code command to read the temperature sensor value

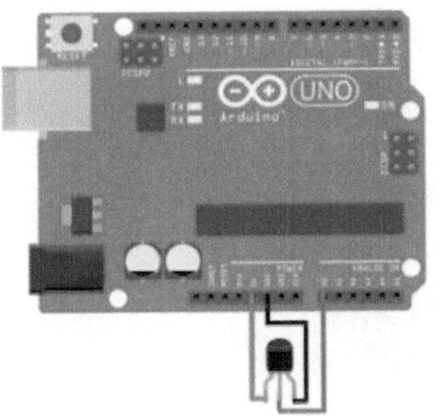
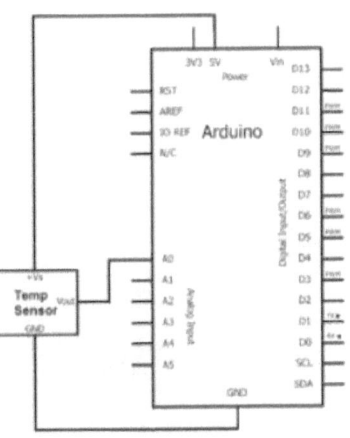

UNIT 3.4
INPUT CIRCUIT SIMULATION

In order to produce input circuit simulation, we can use Tinkercad website for free.

Tinkercad website: www.tinkercad.com

Creating adjustable resistor circuit simulation

1. Add adjustable resistors. Arduino and circuit boards with button "+Components"

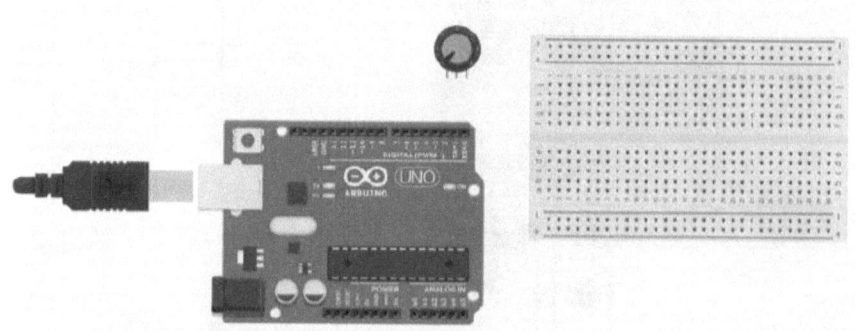

2. Make sure the adjustable resistor value is set correctly by selecting the adjustable resistor and changing its resistance value to 10k or 100k.

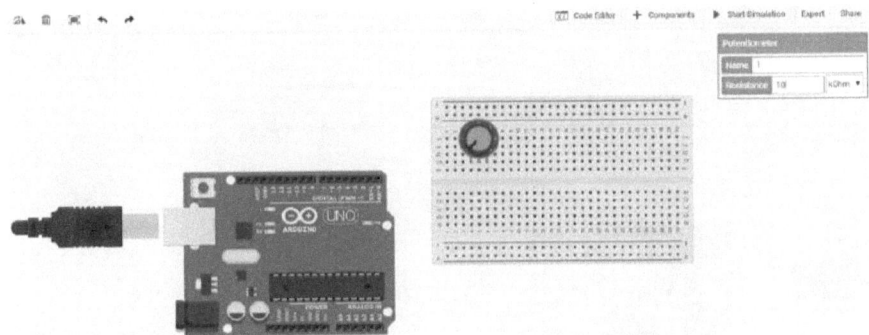

3. Connect the device to the board using the "drag and drop" method.

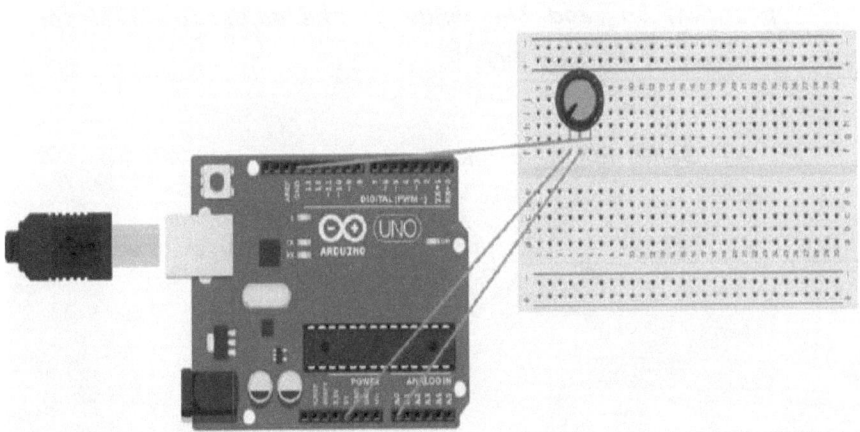

4. Write the program in the "Code Editor" section

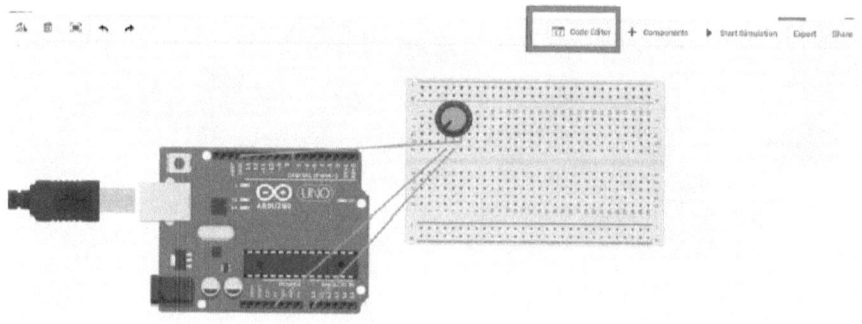

5. "Drag and Drop" the appropriate block to generate a program to read the value of the adjustable resistor connected to pin A0.

a) Generate new variables

b) Set the variable's value to A0 pin reading value.

c) Display reading values with Serial communication (combination of Output block and Variables block)

6. After the program I was ready prepared, press Upload and Run.

7. Serial Monitor "can be simulated with the" Serial Monitor "button.

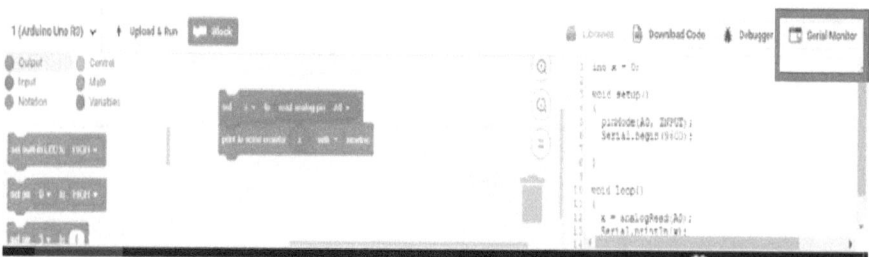

8. When we rotate the adjustable lever, the display value in "Serial Monitor" also changes.

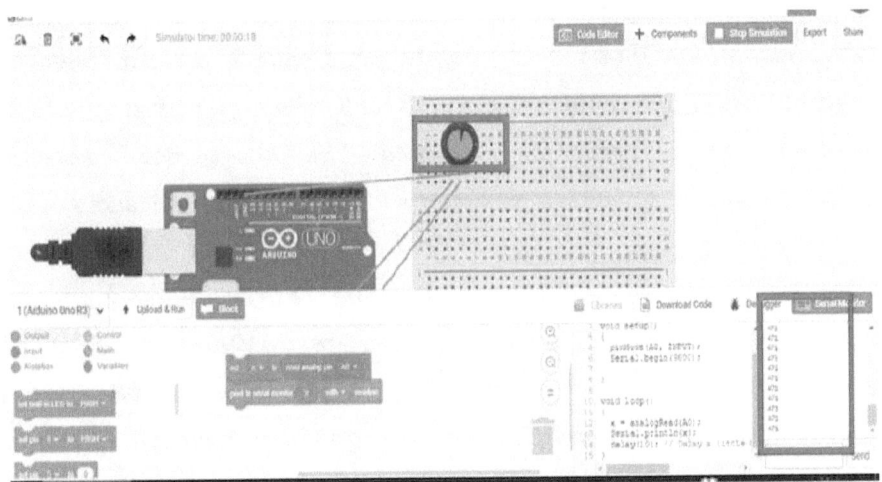

UNIT 4
COMBINATION OF INPUT / OUTPUT

Introduction to Optional Control Programming Structure

We can combine input and output to create a project that can interact with the environment. Recall that the microprocessor functions as a brain to produce the appropriate response (output) based on the input gained (information received).

Using an input device, we can use the read value and compare it to the value we want. If the read value matches the required pattern, the microprocessor can produce the desired output. This is referred to as the optional control structure. The representation in the flow chart is as follows:

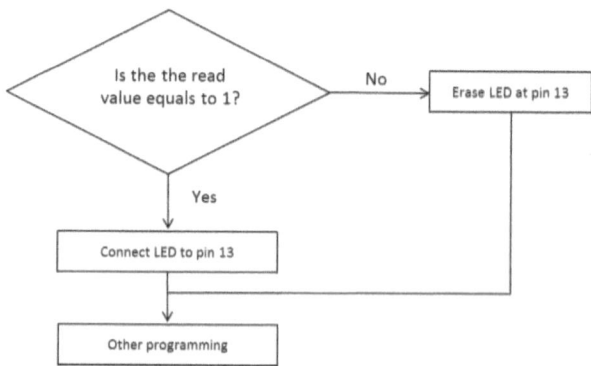

Figure 4.1 (a): Flowchart for optional control structure

The program for the flow chart in Fig. 4.1 (a) is as follows:

```
 6 void loop() {
 7     // put your main code here,
 8     int x = digitalRead(2);
 9     if (x == 1) {
10         digitalWrite(13, HIGH);
11     } else {
12         digitalWrite(13, LOW);
13     }
14 }
```

The comparison operator is used to compare two values and will return true values if the comparison is true.

There are 6 types of comparison operators in Arduino programming: (See next pages for table of comparison opertors).

73

Comparison operator	Function	Example

==	Equals to	x =digitalRead (5);
		if (x == 1) {
		   digitalWrite (13, HIGH);
		} else {
		   digitalWrite (13, LOW);
		}
		
		Read the value of x (switch on pin 5). If x is 1, which when switch is pressed, then a digital signal is sent to turn on the LED on pin 13. Otherwise, send a digital signal to turn off the LED on pin 13.

!=	Not equal to	x =digitalRead (5);
		if (x != 1) {
		   digitalWrite (13, HIGH);
		} else {
		   digitalWrite (13, LOW);
		}
		
		Read the x value (switch on pin 5). If x is not equal to 1, that is, the switch is not pressed, then a digital signal is sent to turn on the LED on pin 13. If the switch is pressed (value is 1), a digital signal is sent to turn off the LED on pin 13.

>=	Greater than or equal to	x = analogRead (A0);
		if (x> = 100){
		　　digitalWrite (13, HIGH);
		} else {
		　　digitalWrite (13, LOW);
		}
		
		Read the value of x, which is the analog input device connected to A0 pin. If the read value is 100 or greater than 100, turn on the LED on pin 13. Otherwise, turn off the LED on pin 13.

>	Greater than	x = analogRead (A0);
		if (x> 100) {
		　　digitalWrite (13, HIGH);
		} else {
		　　digitalWrite (13, LOW);
		}
		
		Read the value of x which is the analog input device connected to the A0 pin. If the read value is greater than 100 (not including value 100), turn the LED on pin 13. Otherwise, turn off the LED on pin 13.

<=	Smaller than or equal to	x = analogRead (A0);
		if (x <= 100) {
		    digitalWrite (13, HIGH);
		} else {
		    digitalWrite (13, LOW);
		}
		
		Read value x which is an analog input device connected to an A0 pin. If the read value is 100 or less than 100, turn on the LED on pin 13. Otherwise, turn off the LED on pin 13.

<	Smaller than	x = analogRead (A0);
		if (x < 100) {
		    digitalWrite (13, HIGH);
		} else {
		    digitalWrite (13, LOW);
		}
		
		Read the value of x which is the analog input device connected to the A0 pin. If the value read is less than 100 (not including the value 100), turn on the LED on pin 13. If not, then turn off the LED on pin 13

To get a value between two values, we need to use the "AND" logical operator.

For example, a value between 500 and 700 can be detected by the program means the following:

```
 6 void loop() {
 7     // put your main code here,
 8     int x = analogRead(A2);
 9     if (x > 500 && x < 700){
10         digitalWrite(13, HIGH);
11     } else {
12         digitalWrite(13, LOW);
13     }
14 }
```

(AND Symbol is &&)

*Example 1 : Especially
for Maker UNO*

The complete code example below is to read the switch connected to pin 2 (for Maker UNO), displaying the value, and if the switch is pressed, turn on the LED on pin 13. If the switch is not pressed, turn off the LED.

(For other Arduino's, please connect the circuit board and convert INPUT_PULLUP to INPUT)

```
sketch_jan07a §
1  void setup() {
2
3      pinMode(2, INPUT_PULLUP);
4      pinMode(13, OUTPUT);
5      Serial.begin(9600);
6  }
7
8  void loop() {
9
10     int x = digitalRead(2);
11     Serial.println(x);
12     if (x == 1){
13         digitalWrite(13, HIGH);
14     } else {
15         digitalWrite(13, LOW);
16     }
17 }
```

*Example 2 : Connection
to circuit board is required*

This example reads values on a light-sensitive resistor, displaying and producing different outputs depending on the read value. When the environment description is dark (read value equal to or less than 400), the LED on pin 13 will be lit. If the condition becomes bright (read value is greater than 400), the LED on pin 13 will be turned off

```
sketch_jan07a§
1  void setup() {
2
3     pinMode(A0, INPUT);
4     pinMode(13, OUTPUT);
5     Serial.begin(9600);
6  }
7
8  void loop() {
9
10    int x = analogRead(A0);
11    Serial.println(x);
12    if(x <= 400){
13       digitalWrite(13, HIGH);
14    } else {
15       digitalWrite(13, LOW);
16    }
17 }
```

UNIT 4.1
COMBINATION OF INPUT CIRCUIT TO OUTPUT CIRCUIT

Input and output circuit connection does not differ much from the input circuit only or the output circuit connectivity only. The following are some examples of input and output circuit.

Circuits that include light sensitive sensors, buzzer and LED

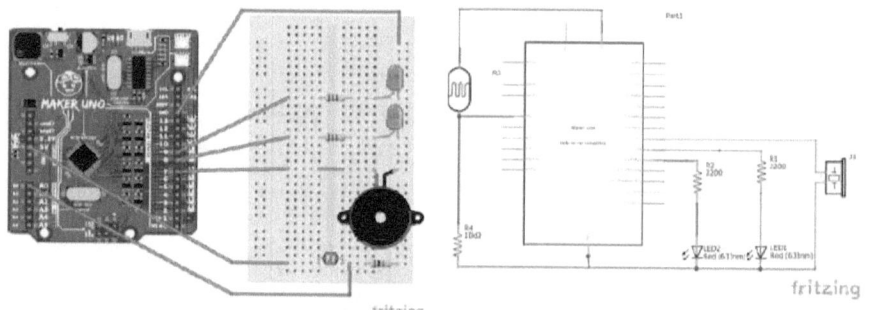

With this connection, we can produce one simple music box. When the box is open (brightness increases), the microprocessor will play music and turn on the LED.

Circuit that include switch, buzzer and LED

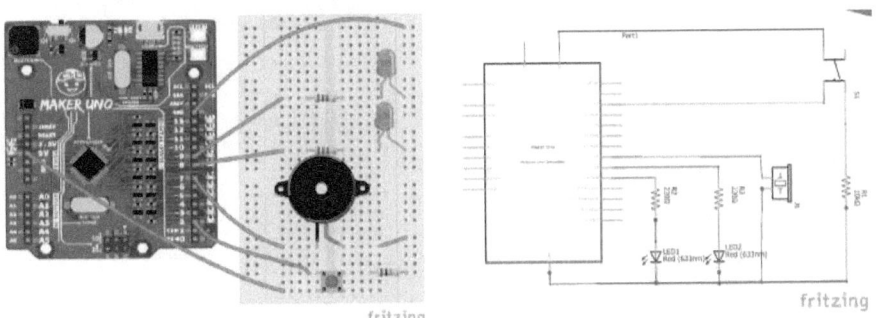

With this connection, we can produce a music box activated by the switch. When the switch is pressed, the microprocessor will play the music and turn on the LED.

UNIT 4.2
INPUT AND OUTPUT SIMULATION

To generate Input and output simulations, we can use the Tinkercad website for free.

Tinkercad website: www.tinkercad.com

A combination of input and output circuits involving switches and LEDs. When the switch is pressed, the LED will turn on.

1. Add switch, resistors, LED, Arduino and even circuit board with "+Components"

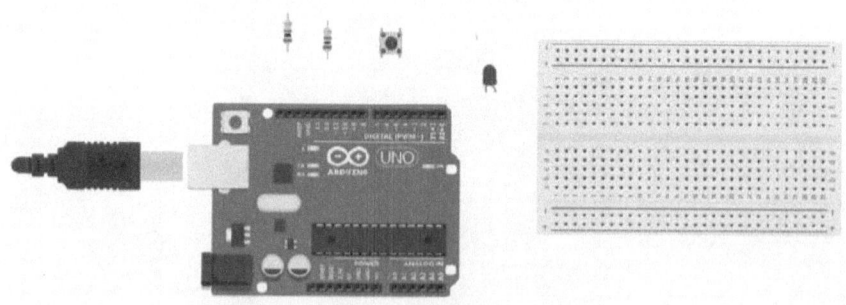

2. Make sure the resistor value is correct by selecting the resistor and setting the value to 220 Ohm and also 10k Ohm

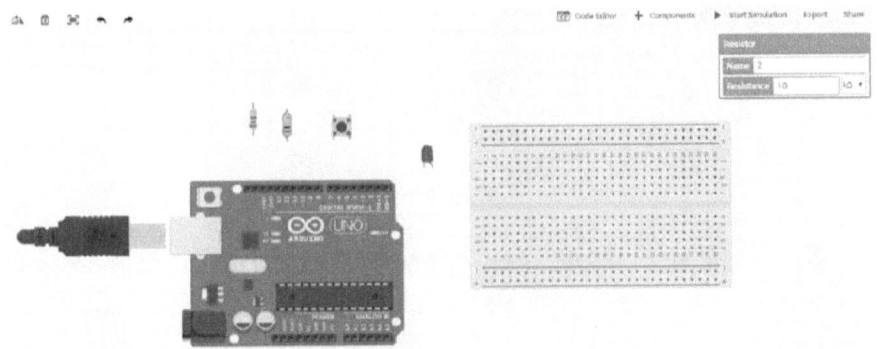

3. Connect the devices to circuit board using the "drag and drop" method.

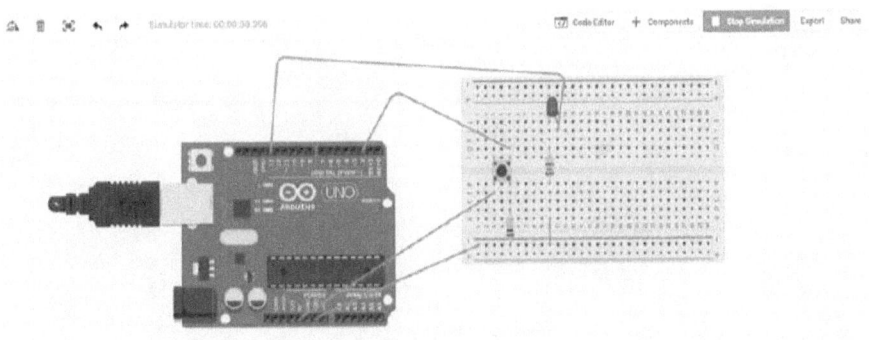

4. Click on "Code Editor" and write a program

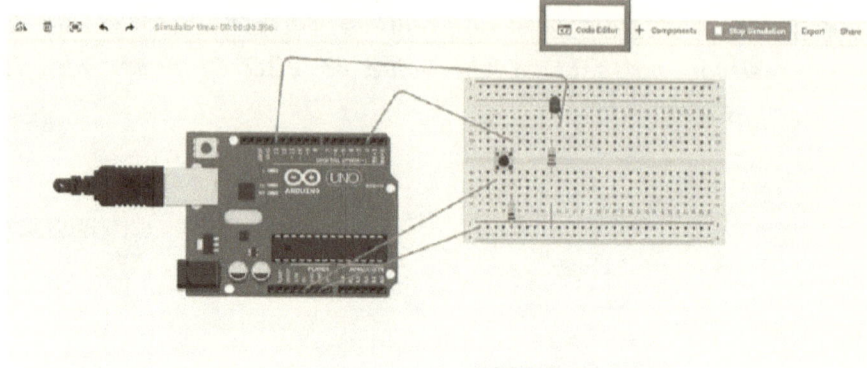

5. "Drag and Drop" the appropriate block for programming to allow the LED to flash on pin 5.

   a) Generate new variables

b) Set the value of the variable to the reading value of pin 2.

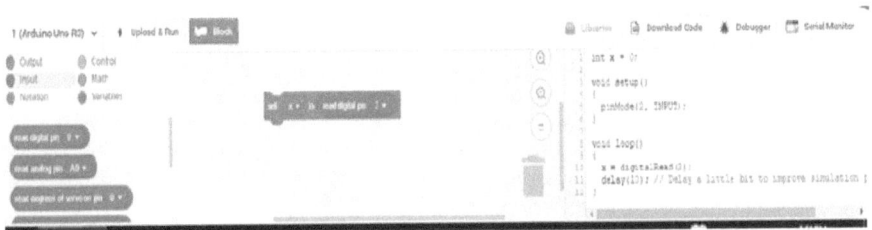

c) Display the reading value with Serial communication (combination of Output block and Variables block).

d) Generate the optional control structure block by selecting the IF ··· THEN ··· ELSE block from the Control block.

Then, use the Math and Variable block to generate the following block combinations:

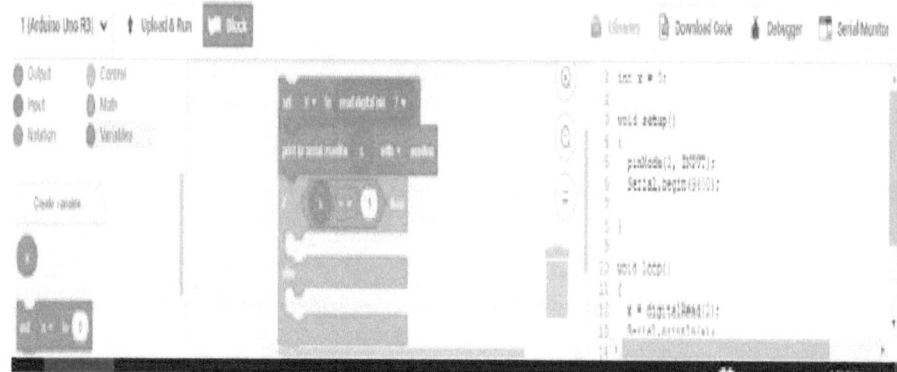

e) Pull the block to turn on and turn off the LED on pin 13 into the optional control structure block.

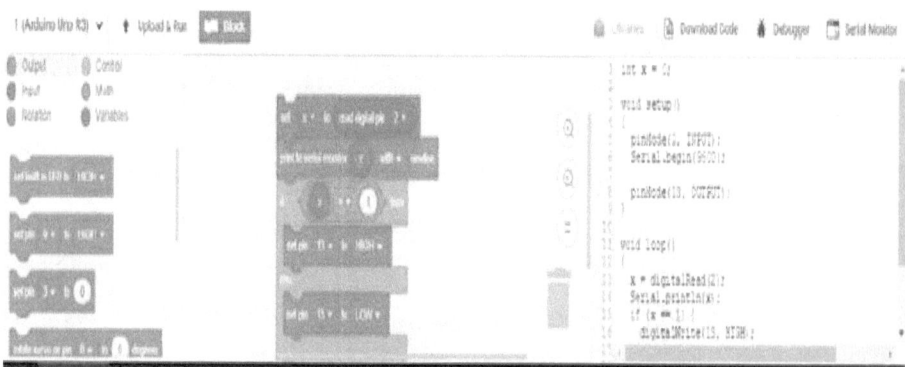

6. After the program was completely prepared, press the Upload and Run.

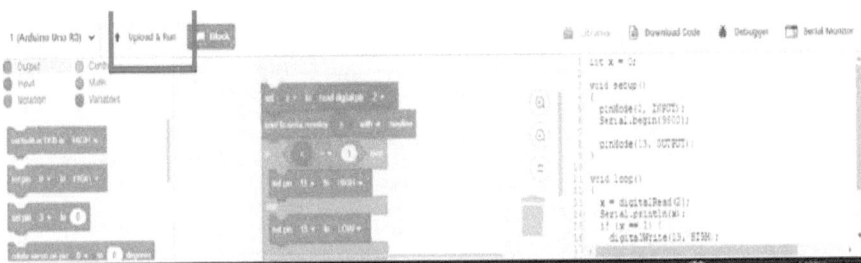

7. "Serial Monitor" can be simulated with "Serial Monitor" button.

8. When the switch is pressed, the LED will turn on. The values in the "Serial Monitor" will change as well

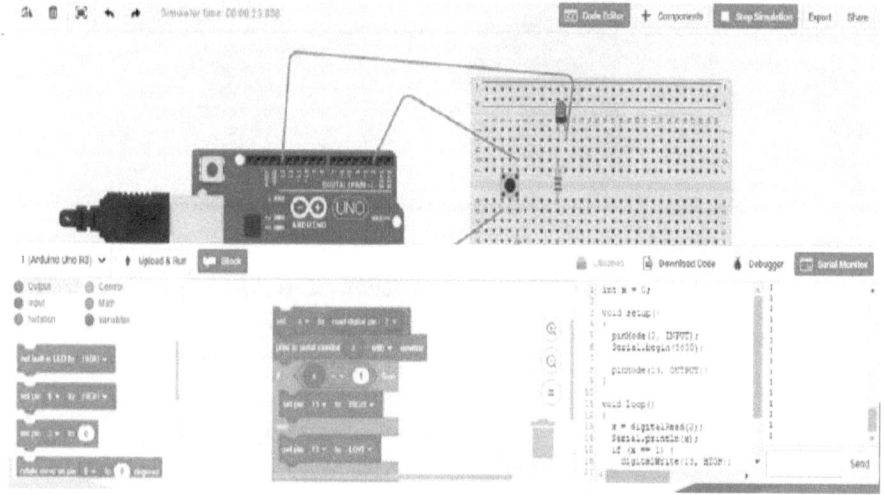

UNIT 4.4
PROJECT EXECUTION

Here are the steps to create a project using microprocessor, output device, input device and even software

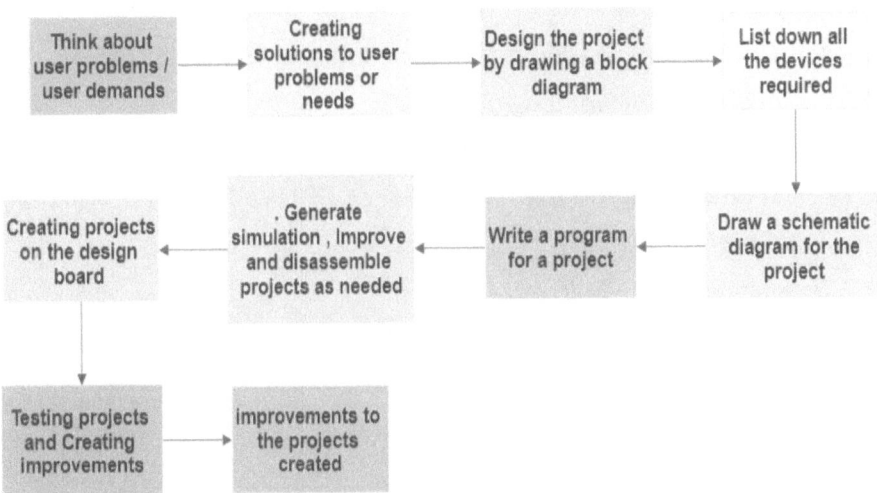

APPENDIX 1

Programming using a Smart phone.

Steps:

1. Determine whether the smartphone has On-The-Go (OTG) functionality.
2. Search the web or search under 'Settings' on your smartphone or use the USB OTG CHECKER application in Google Playstore.
3. Activate OTG if necessary
4. Download the Arduino Droid application on the Google Playstore.
5. Write a program

6. Select the flash icon to compile
7. Select the flash icon to compile. 5. Select Settings—> Board Type -> Arduino -> Uno CH340G (Maker UNO) or Uno (Arduino).

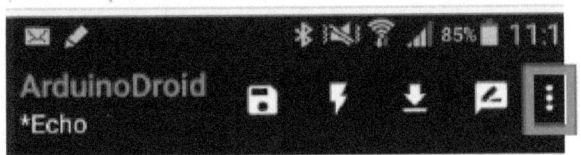

8. Connect the microprocessor to smartphone using a USB cable and OTG cable.

9. Press the upload to send the program how to microcontroller.

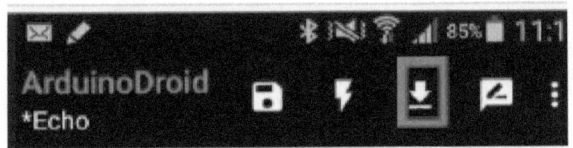

www.ingramcontent.com/pod-product-compliance
Lightning Source LLC
Chambersburg PA
CBHW020601220526
45463CB00006B/2401